Tunnels & Caverns
By
Harry J. Hilton

Looking Back at Forty Years

in

Underground Construction & Estimating

Copyright © 2014 by Harry Hilton.

ISBN: Softcover   978-1-4990-3646-6
              eBook   978-1-4990-3645-9

All rights reserved. No part of this book may be reproduced or transmitted in any form or by any means, electronic or mechanical, including photocopying, recording, or by any information storage and retrieval system, without permission in writing from the copyright owner.

Any people depicted in stock imagery provided by Thinkstock are models, and such images are being used for illustrative purposes only.
Certain stock imagery © Thinkstock.

This book was printed in the United States of America.

Rev. date: 06/28/2014

To order additional copies of this book, contact:
Xlibris LLC
1-888-795-4274
www.Xlibris.com
Orders@Xlibris.com
636169

CONTENTS

Part 1   Hydro-Electric Power Stations 13

Part 2   Tunnel Boring Machines & Drill & Blast 21

Part 3   Specific Project Details 31

Part 4   Estimating 37

Part 5   Encounters With Adverse Ground Conditions 57

Part 6   Pictures & Sketches 67

Part 7   "Murphy's Law" Episodes 77

Cover Picture

Chicago TARP Project
32 Ft. Jarva T.B.M. Tunnel

Picture by H. J. Hilton

DEDICATION

I wish to dedicate this book to my wife Beverley. Only through her incredible resourcefulness and tenacity in moving, with five young children to the isolated sections of Labrador (two separate occasions) and later to a number of urban areas in Canada and the U.S., were we able to maintain our close family relationship. Of course I should also include sons Barry, Mike & Brendan, and daughters Mary & Eileen, all of whom had so many adjustments to make but learned to be very adaptable !

FOREWORD

The author wishes to point out that this writing does not purport to represent the *design* aspect of any project (although specific dimensions, capacities, etc. are indicated in many instances). The intent is simply to stimulate the *vision* of a student (Civil, or Mining Engineering) to the endless possibilities of utilizing Underground Space, *particularly in the area of (fuel free) hydro-electric power development and other cavern type projects*. It is also hoped that some of the unexpected incidents that are recounted from various projects will be of interest to the tunneling "hands" of today!

****

I also would like to make it clear up front that I am *not* a Graduate Engineer. I started my working life at age eighteen, as a "Trainee Engineer" - to use the term of the day - but found that in very short time (about one year) I was "on my own" on a fairly large project handling all of the Field Engineering - which is primarily the survey /layout, plus making lots of field sketches for Supt./foremen. I did continue with Evening Classes, but with moving from project to project could not maintain a continuity.

However, in many ways I really regret not sticking with the studies. Had I obtained that degree it would have opened doors that I have found were closed to me on occasions. Particularly in this day & age, those papers mean so much!

GLOSSARY OF TUNNEL TERMS

TBM	*Tunnel Boring Machine (Rock)*
Mole	*Boring machine for "soft ground"*
Soft Ground	*Soil to the firmness of soft shale*
Primary Supports	*The initial arch supports during excavation of tunnel*
Horseshoe Tunnel	*Sides & Arch form a horseshoe shape, invert (bottom) is flat*
Modified Horseshoe	*Sidewalls are straight, arch curved*
Lagging	*Hardwood timbers placed behind or between steel ribs (sets)*
Shoulder Crown	*Point where vert. wall meets the arch Top point of arch (Overt)*
Catenary Line	*Overhead El. power line . . . Trolley-Line*
Rock Bolts	*Bolts designed specifically for rock support*
Steel Sets	*(Steel Ribs) circular or U shaped steel beams shaped to tunnel diameter*
JV	*Joint-Venture, two or more contractors form an alliance to bid a project*
JV Sponsor	*The partner who will have the major part of the investment and provide most, or all of the project personnel*

A Couple of Definitions

In the context of this writing, there are two distinct areas of underground excavation:

TUNNELS: Strictly linear, excavated by machine, or by hand.

CAVERNS: Underground chambers, usually excavated by drill & blast methods.

Combinations of the two have been used many times, and in fact the driving force behind the writing of this little book, is the hope of encouraging interest, by students and engineers, in the application of this *combination* feature. This I truly believe would lead to further design and development of Underground Power Stations, and thereby, expand production of the cleanest and safest form of electrical power available.

The other aspect of this writing is to simply demonstrate some of the day-to-day aspects of "life in the field" *and* in the world of Estimating, especially estimating overseas projects - which literally can be a whole different world!

In instances where I have had direct involvement with certain projects, I have included details of construction methods, equipment used, and various problems encountered.

There is a large amount of information on the Internet relative to this entire subject. For example, on researching the history of the Niagara Falls Generating Stations, I discovered, to my surprise, that dating back to the 1800's a *number* of various types and sizes of power stations, underground and above ground, were constructed in the Niagara area - both on U.S. and Canadian sides. The history of the area over a period of a hundred years provides fascinating reading and insight. This narrative however, concentrates more on the "hands-on" aspect of construction.

Part 1

Hydro-Electric Power Stations

Types of Hydro Electric Power Stations

The following definitions are used to clarify the difference (at least from this writer's viewpoint) between what may be considered conventional power stations and underground power stations.

There are actually two types of *conventional* hydro-electric power-stations:

1. Utilizing a Natural Waterfall

A site, just downstream of the waterfall is selected and a power station is constructed adjacent to the river. The river is partially diverted a short distance *upstream* from the waterfall, through man-made penstocks (large pipes in effect), and directed at a point some short distance behind the powerhouse. From that location an elbow section and a short horizontal section of penstock will carry the water to the turbine scroll case. The force of the water hitting the turbines is converted via the turbines and generators to electricity. Example: Niagara Falls, both U.S. and Canadian sides. There is no minimum specified height for a dam or penstock, it's really a question of determining the minimum power required and relating that to an available source, and height of waterfall. For instance, there is a small power development on Vancouver Island where the water is brought from a lake about two miles away, through a 12ft. (3.6m) diameter tunnel. The tunnel terminates at a hillside just behind the powerhouse at an elevation of a hundred feet or so above the floor of the powerhouse. From that point the water is dropped through three steel pipes into the generator locations. The total fall from the lake intake, to the powerhouse floor, may only be out 200 ft. (68m), but apparently the power generated was worth the investment. Outlet pipes from the powerhouse discharge the water to a nearby river, which is obviously at least at 200 FT (61m) below the elevation of the lake - at the intake location.

2. **Man-made Dams**

The second form of conventional power-station would seem to be a logical progression from the natural waterfall type. "If we build a dam across one of our huge rivers, we could then draw from the overflow to simulate a waterfall, install penstocks, build a powerhouse on the downstream bank, and we are in business". One small item - in order to build the dam we have to temporarily divert the river and create a dry area in which to work. No problem, simply install a diversion tunnel(s) and temporarily relocate the course of the river over a distance of a few thousand feet. Great example—Hoover Dam and Generating Station.

Underground Power Stations

This particular application of Topography is emphasized in an attempt to promote serious investigation into the possible existence of yet undiscovered sites for such projects throughout the U.S. Granted, initial costs are high, but once the power-house structure is in place, power is generated at minimal cost, utilizing only natural water - with hydrostatic head - to spin massive turbines and produce thousands of megawatts of power.

So, we must find an area where the natural topography would in itself provide a head of water, and of course must also provide an egress for the water flowing *from* the turbines. There were a number of underground power-stations in existence long before the 1960's, but I will reference only latter era projects - in which I had some involvement.

There are two projects that come to mind when I think of the topography factor. One is the Churchill Falls Generating Station in Labrador, Canada. The second is the Manapouri Power Station in southern New Zealand.

The Churchill Falls Project

An existing waterfall, located on the Churchill River in Labrador, Canada was the first feature that prompted surveying and speculation into the possibility of developing a large power project in the area. Of course it was obvious that a two hundred and fifty foot height would not justify development in such a remote area. However, study of the surrounding area, of some five miles radius from the waterfall, indicated that an underground powerhouse could be built at up to *a thousand feet below ground,* and still have a natural slope to carry the "used" water - through tailrace tunnels - back to the original river bed, some few miles downstream from the falls.

Following a number of years of surveying, estimating, "politicking", and bold resolve, the project was underway.

The Churchill Power Development is capable of producing 650,000 KW. The major portion of this power is sold to the Provinces of Newfoundland and Quebec, but some also goes to the New York State Grid.

The "Manapouri Power Development"

This project was started in the early 1960's. It is located on Lake Manapouri which is in an area known as Fiordland National Park, at the south end of the south island of New Zealand. The general area borders on a fiord of the Caspian Sea called Doubtful Sound.

The Manapouri lake surface is at an elevation of some 200 meters above sea level. Some entrepreneurial people figured that a powerhouse could be located *below* the lake, but still *above* sea level, with a vertical drop of almost 200 meters *and* a means of discharging the water, to the ocean, by simply providing an outlet tailrace tunnel. Well, it became a reality.

Pumped Storage

Another variation of Underground Hydro Power is "Pumped Storage". Actually this system is quite possibly the most common type of hydro power stations worldwide. There are many such systems in the U.S. One is in California, on the Feather River near the town of Oroville. Both the Dam and the Powerhouse were part of what was called "The California State Water Project". The overall plan, to bring water from Northern California, to the southern region, for both irrigation and drinking water supply, involved many other tunnel, pipeline, and reservoir projects. At that time (1963/64) the total estimated cost was $439 million.

The Oroville dam, I believe, is still the largest dam in the U.S. Built in the sixties, the dam is composed of earth fill, rock armor, and a *relatively* small concrete core. It is 735 ft. high and contains something in excess of eighty million cubic yards of fill material. (See Part 4 for more information on the Oroville Dam).

Back to Pumped Storage: Simply put, the powerhouse is constructed underground, but there is no natural fall below the turbine elevation for egress of the water. So, a containment area of sufficient volume to store the water from twelve hours of operation is excavated underground. All of the power generated is connected with a Surface Grid. At off-peak hours (night-time) the combination pump/turbines are started up and the stored water is pumped back up to the lake. The cycle repeats every twenty-four hours - another means of generating electricity from falling water. The power used to pump the water back to the lake is purchased from the "grid" consortium at a fraction of the cost of the day-time power, and of course is still generated by water.

Another great example of pumped storage is the Helms project in Colorado. This huge development includes some of the deepest shafts used for such a project.

On the subject of the Storage Area, I have to emphasize the importance of studying the use of TBM's to develop areas that in the past may not have been considered feasible or economical. This is that "Combination" factor that I believe needs more exploring.

Part 2

Tunnel Boring Machines & Drill & Blast

Rock Boring Machines - TBMs

The advent of the Tunnel Boring Machine (TBM) or specifically the Rock Boring Machine was a huge milestone for the world of tunneling. To have a machine that could bore a tunnel to a specified diameter, producing a near-smooth rock surface, at a rate of four or five times that of a drill-blast operation and with a *much higher safety factor,* what a wonderful advancement in technology, and what a boom for contractors and engineers.

In the early development days, machines in the twelve to fourteen foot diameter range were probably the most common. All TBMs are powered by electric motors. The motors are located at points around the perimeter of the cutter head. Hydraulics pumps and electric motors combine to power the cutter head. A 14 ft. (4.3 m) machine might be powered by four forty h.p. motors. Over the years, motors up to 100 h.p. are used. A 32 ft. (10 m) diameter TBM would utilize perhaps twenty-four motors. The actual cutting of the rock is accomplished by hardened steel cutter wheels each approximately 13" (0.33 m) in diameter. These wheels are strategically positioned around a circular steel plate - the Cutter Head. The placement of the cutters will result in a series of two inch deep "Kerfs" - V grooves at about four inch centers around the full circle of the wheel. As the cutter-head is pushed ahead on its mounting beam, the sheer weight of the TBM will crush the high point of the kerf, creating a mass of crushed rock. It is Interesting to note that the *spoil will actually "bulk" as much as one hundred percent,* meaning, every solid cubic yard of rock excavated, will measure two cubic yards in the spoil heap. The cutter-head of a thirty foot diameter machine would contain eighty or more roller cutters. By testing sample drill cores of the rock in the area of a proposed tunnel, the manufacturers of these cutters can predict, quite accurately, the life of the cutters. This is a great help for estimating the cost of cutters for a large project. At thirty feet diameter, twenty-six cubic yards of solid rock will be mined for every linear foot of tunnel advance. Also, the cutter wheels will require periodic replacing, a considerable cost factor in the estimate.

Of course, we are well aware that for every action there is an equal and opposite *reaction*. The rock TBM and the soft-ground machine - also known as the "Mole" - differ in this regard. The TBM utilizes side-grippers to exert lateral pressure against the surrounding rock. These are hydraulically operated, arc-shaped steel sections, conforming to the tunnel diameter. There may be four or eight grippers, depending on machine diameter. The contact surface of the gripper will be in the order of three feet wide by five feet arc length. Of course there will be as many variations in size as there are machine sizes. Once the operator has tightened the grippers against the side walls, power can be exerted to thrust the body of the TBM ahead again on its travel beam for the "stroke" distance - four or six feet (1.2 or 1.8 m) depending on machine design. The TBM is now ready again to start grinding rock.

Today's TBM's can be extremely sophisticated. A large TBM will contain an impressive command-post, giving the operator finite control for steering, pressure exerted, installation of huge precast concrete support segments, plus an array of sensing instruments. One of these sensors is a unit that will detect the presence of combustible gas, at the tunnel face - at a very small percentage of total air. The sensor will immediately trip a breaker, causing the electrical power to the TBM to shut down. I can vouch for the effectiveness of that little gadget, having witnessed a number of such shut-downs in a tunnel below Lake Ontario. Usually, such interruptions were of short duration, as the gas - methane in this case - would dissipate as soon as it mingles with the forced air coming from the tunnel ventilation line. Meanwhile, of course, the tunnel would be evacuated until the exiting air was tested again and found to be clear.

Soft Ground Machines - "Moles"

The soft ground TBM is pushed ahead by hydraulic "push-jacks". These again will vary in number depending on the diameter of the machine. The ground support system is built inside the shell of the machine. As soon as the TBM has been buried, e.g. pushed ahead to a point where the back end of the machine skin is at the edge of the access (construction) shaft, the support system - either ribs & lagging or concrete segments - is installed inside the tail section of the machine. The push-jacks now thrust off the last support set, and so the cycle continues in four foot (1.2 m) pushes.

In the earlier machines the cutter-head was a large open wheel with spokes. The spokes were the cutter blades, peeling the dirt from the tunnel face, quite effective as long as the soil was reasonably stable.

Prior to the introduction of the Mole machine there was a form of mechanical protection, used by the sand-hogs, called a Shield. This device was in effect a steel cylinder built to perhaps a couple of inches larger diameter than required tunnel external diameter, the overall length somewhere between six and ten feet. At the front end, the top half of the circle was tongued to project out in front of the full diameter providing some protection for the miners against soil collapses ahead of the main shield - a quite common occurrence in very soft ground. In wet sandy material it was sometimes extremely unsafe or just impossible to mine even with the protection of a shield. If a major project - usually sewer or waterline tunnel - had to be driven through such material, it was necessary to resort to the use of compressed-air, in conjunction with a Shield. It was quite amazing how the application of air pressure, as little as five p.s.i. would just dry out very wet sand. In effect the air was literally pushing the water out of the sand. *The pressure required is directly relative to the height of the water-table above the tunnel.* Water pressure equates to 2.45 ft. per pound. Ten feet head of ground water would require five pounds of air pressure to maintain a dry tunnel, allowing a margin of error. The real trick is not to push the water too far out, as the sand may become *too dry* and tend to fall!

Another form of dewatering that is quite effective is the "Eductor" system. This system entails the installation of vacuum operated suction pipes, usually just two inches in diameter, but placed at twenty feet centers along the route of the tunnel. The suction pipes discharge into a 6" header line, which will run along the route of the tunnel. Pumping will be activated for some days before the tunnel reaches a given point.

Since the 1980's a number of companies have been manufacturing machines that incorporate combination type cutter-heads capable of excavating through a variety of wet soils and "mixed-face" material. Cutter-heads might be "Earth Pressure Balance" (EPB), Slurry Injected, or may contain localized compressed-air chambers. The spokes were replaced with solid steel plates shaped to cover the full circle and made in tapered sections that could be "fanned" open and closed by the operator. Some Moles may include a rock-crusher as it is not totally unusual to encounter random boulders in certain types of soft soils. These machines may be custom-made for a specific project. They would obviously be quite expensive and only warranted for a considerable length of tunnel, such as a section of subway or large diameter sewers or waterlines.

Roadheaders

Roadheaders are machines designed to cut rock in "irregular" cross sections, i.e. shapes that are not circular by design - and usually of relatively short length, where installation of a T.B.M. may not be economical. These machines are quite efficient in cutting rock up to a limited compressive strength. They are physically quite large, somewhat similar to a backhoe, and therefore require an entryway of considerable size. The Roadheader is caterpillar- track mounted, has a solid boom which supports a rather small rotating cutter-head which cuts a swath of some inches in depth. Simply put, the tunnel face is advanced as these swaths eventually cover the total face area - to whatever final shape is required. This is very useful when irregular shapes are required, but obviously, this machine will only cut rock in the lower compressive strength range.

There are a couple of other methods of tunneling that I should mention. One is a system that was basically developed for excavation of large underground openings in soft or mixed-face ground. It is generally known as "The New Austrian Tunneling Method"

It incorporates extensive study of the geology of materials to be excavated, and the application of soil mechanics. Support systems may include steel ribs, shotcrete, wire mesh, lattice girders, etc. or any combination thereof. This system has become quite common in recent years for specific type projects. Boring machines however are not usually used in conjunction with this system, at least to my knowledge.

****

The other unusual method - used mainly for installing a short tunnel - under a highway for example - utilizes steel pipes or even mini tunnels as support for a very large diameter opening in soft ground. This system is referred to as "stacked tunnels".

There is one major project that I know of where stacked tunnels were used to create a fifty foot diameter (15.2m) tunnel in soft ground. This is on a section of highway 90, in Washington State, known as the Mount Baker Ridge Tunnel. I prepared a cost estimate for construction of the tunnel in 1984.

Drill & Blast Tunnels

Before the era of TBM's, drill and blast was the only method of underground rock excavation - that is of course if we don't consider the various types of coal mining machines - but that's different territory.

D & B goes back a long way, probably to the early days of gunpowder. Just make some holes in the rock, pour some powder in, light a fuse and BOOM!

Of course there were no pneumatic drills back then so somebody figured that hammering on a heavy steel chisel/bar, rotating it at the same time, would cut a hole of some depth into the rock. A lot of tunnels were driven that way for the original North American railways, U.S.A. and Canada.

Inevitably the "jackleg" drill evolved and life surely became a lot easier for those miners. Jump ahead a few years and we now have "drifter's drills" - much heavier machines mounted of "feed shells".

The feed shell was in effect a steel beam, or channel, mounted on a hydraulic boom. The feed mechanism is located inside the channel. There were two types of feeds, a chain feed and a screw feed. The actual drill is set on top of the feed, fixed in position and is powered ahead when the miner activates the drive motor. The miner has total control of the drill speed. The speed of course is relative to the hardness of the rock. This is where the skill of the miner is so important. In hard abrasive rock the miner must check the condition of the drill-bit after each hole is drilled. In very abrasive rock the bit inserts can wear very quickly.

Anyone who has worked in D & B tunnels knows that the smaller the diameter of the tunnel, the higher will be the powder requirement - <u>per cubic yard/ meter</u>.

Drill patterns for different size tunnels will vary considerably. (All of the above of course, assumes similar rock types) Therefore, it is obvious that a larger number of holes - per square meter - will be required in the smaller tunnel in order to insert sufficient dynamite for the blast.

Perimeter Holes: There are many factors in tunnel blasting but one item I would stress here is safety. In connection with safety, I suggest that the use of "perimeter holes", in the tunnel arch, is highly recommended.

Perimeter holes are smaller in diameter than blast holes, but drilled much closer together. There is a special powder available for these holes that is only one half inch (1.3cm) in diameter. This powder is designed to shear *between* holes and thereby reduces "overbreak" and minimizes shattering outside the required perimeter.

Bootleg Holes. These are short pieces of holes that are usually left in the tunnel face after a round has been blasted. There might be just a few, there could be many, or none at all! For example, if a round of eighty holes is drilled at say 12 ft. (3.65m) deep by each of the miners, and the average depth broken is 11ft. then there is one ft. of bootleg left. This *may* be acceptable, but if it tends to increase, there is something that needs to be changed. Usually it's the drill pattern, and most likely it's the "cut" or center area holes that require adjusting. A good Superintendent or Shifter will figure where to start changing the pattern, adding to the number of holes etc.

Of course changes in the type of rock being drilled will have an effect on the type of break obtained—for a given drill pattern and amount of explosives.

On our rail tunnel in Labrador our North heading was in a very brittle granitic rock with a high Quartz content. Extremely hard on drill bits, but broke to full depth almost every time. The south heading however, for a few thousand meters was through a much softer, almost spongy rock* that drilled very easily but we found after a few rounds of drilling twelve feet (3.6m) we were only breaking an average of 9.5 ft. (2.9m). This was not acceptable! We made adjustments to the "cut" area, adding one extra "burn" hole, 4" diam., plus an extra "delay" in the firing sequence. This improved the break depth to an acceptable degree. At this point the "boss" has to make the call—just figure the relative costs, does the extra depth broken compensate for the extra time spent drilling the "cut" (maybe twenty minutes), not hard to figure !

Note: I have heard stories of miners drilling into bootleg holes—to save some time. Unfortunately holes could contain a small amount of powder which was actually left intact and could blast enough rock from the collar to seriously injure the miner. Just the friction of the drill bit hitting the powder fragments set the minute blast off, and that was enough to seriously the unfortunate minor. Hopefully that practice has been stamped out now. Would recommend however, especially in places where miners may be inexperienced, that *very severe* warnings be given to crews - at all Safety Meetings - regarding those Bootleg holes!

*Mica Schist.
At the north heading the rock body was a very abrasive, high quartz content.
Granite.

Part 3

Specific Project Details

The Chicago TARP Project

I made a brief reference to the Channel Tunnel, a truly outstanding engineering, and construction achievement. There are however, some surprisingly large tunnel projects, some completed, some still under construction throughout the U.S.

A particularly notable undertaking is known as the TARP project, in the Chicago, Illinois area. The acronym stands for Tunnel and Reservoir Plan, (long before it was used in the banking business!) Simply put, the Plan was developed to control (a) the severe flooding caused by rainwater, and (b) the amounts of raw sewage that was overflowing into Lake Michigan, and local rivers, as the existing treatment plants and sewers were simply unable to handle the volume.

Extensive studies were made to assess the feasibility of building additional and larger, treatment plants, but the scales finally tipped in favor of the Reservoir Plan.

The main tunnels run from the South side of Chicago to the Calumet area. I believe the total lengths, including branches, is in the order of some 109 miles (174 Km). The main line includes perhaps 40 miles of 30 ft. to 35 ft. (9m to 10.6m) diameter tunnel. In order to position the tunnel in the ideal material for excavation, i.e. competent limestone, it appears the depth of the tunnel would have to be about 250 ft. (76m) below surface. Of course that may also have also been necessary in order to obtain a continuous downgrade slope from the far reaches of the tunnel(s).

Overall this was a hugely complicated series of projects which led to bold undertakings on the part of the engineers and the contractors. Not only were TBM's of sizes larger than ever built before required, but systems for placing the concrete linings while the mining still progressed required more innovations. At an average of 31 ft. diameter, the concrete used would be approximately five cubic yards per lin. ft. of tunnel. 500 cubic yards per day would not have been uncommon. This also entailed moving and resetting 120 L.F. of thirty foot diameter steel form each day.

My personal involvement in the TARP undertaking was mainly cost estimating*, plus occasional site visits. However, the tunneling contractor I worked for (ten years as Chief Estimator) in Ohio, S & M Constructors, also had a subsidiary manufacturing company named "Jarva". The founders of this company, the Scaravilli family, were pioneers in the development of TBM's - and the use thereof. Jarva built a 35 ft. diameter machine and a number of smaller machines for the TARP system.

* See the section on "Estimating" for further item on the TARP system.

Churchill Falls Underground Powerhouse

In my forty plus years, working specifically for tunneling contractors, I was involved to some degree or other in the cost-estimating and/or construction of a number of Underground Powerhouses. However, by far the most intriguing, and demanding, project, of any type, that I was involved with, (two years on site as Chief Field Engineer) was the "Churchill Falls Power Development", in Labrador, Canada. This mammoth power station, built totally underground, produces some 5,428 mega-watts, or 7.2 million h.p., utilizing eleven generators, located some 1,000 ft. below surface. (By comparison, Hoover Dam produces about 3.0 million h.p.). Why was this huge power development located in the barren wastes of Labrador? Simple, the specific site possessed the topographical features, plus the availability of huge volumes of water. By building *relatively* small diversion structures, and some forty miles of dykes, (average height only twenty feet or so), a huge man-made lake was formed. From there the water was dropped *twelve hundred feet* (366m) - at an angle of 70 degrees - through eleven twenty foot diameter (6.1m) concrete lined penstocks—to the underground generators.

The fact that the powerhouse was located in such a remote location did present many problems. Transmission of power at 750 KVA entailed for one thing, the design of equipment, including insulators larger than any available at that time, but the sheer volume of power generated justifies the cost.

The volume of rock excavated to develop the myriad of caverns forming the underground complex, was just staggering—about 2.5 million cubic yards.

> Powerhouse Chm.: 1200'L, 100' H, 80' W. (366 x 30 x 24m)
> Surge Chamber: 800' L, 150' H, 80' W (244 x 46 x 24m)
>
> Transformer Gallery: 1,000' L, 30' H (395m x 9 m)
>
> Main Access Tunnel: 28' H, 33' W, 5000' L.
>   (8.5m x10m x 1,524m)
> Tailrace Tunnels: (2) 60'H x 45'W x 5,000' L.
>   (18.3 m x 14m x 1,524m long)
> Penstocks (11): 25 ft (7.6m) exc. dia. x 1,200 ft.(366 m)

Connecting tunnels, draft tube chambers, etc. accounted for another large percentage of the excavation.

All of the concrete was mixed at the large on-site batch plant, hauled underground in Ready-Mix trucks, and distributed from the East Forebay to required locations, by a combination of pumping and overhead cranes. Throughout most of the caverns and tunnels, rock-bolts, varying from 15 to 20 ft (6m) in length,* were used as the main rock support for the arches and walls. Also, throughout all arch areas, welded wire fabric was installed in conjunction with the rock-bolts.

* Bolts 2" in. (5 cm.) diameter, up to 75 ft. (23m) long (coupled in twenty and fifteen foot sections) were used In the Surge Chamber walls where a geological fault was encountered, between the Power Chamber and the Surge Chamber.

Mount Baker Ridge Tunnel

One of the largest soft ground tunnels ever built, is a section of highway tunnel approximately 1,330 ft. long, known as the "Mount Baker Ridge Tunnel". (I prepared an estimate for this project, but did not actually see the work performed, as we were not the successful bidder).

This very unique project design called for a total of twenty-four tunnels, each nine feet (+/-) in diameter* stacked together to form a fifty foot (15.2m) inside diameter circle. The small tunnels would in effect become the primary support, *and* the final support for the finished road tunnel. Twenty-three tunnels were excavated by custom built machines, then a "Keystone"" drift was excavated by hand-mining. Precast concrete segments (more horseshoe shaped than circular) were installed inside the excavator for all drifts. As each drift was completed it was immediately filled with concrete before the next drift was started.

On completion of all drifts, bulk excavation, using large front-end loaders and dump-trucks is started, working of course, from the top down. No secondary support was required. When all excavation is complete, installation of the concrete lining, including three levels of roadway proceeds.

* In this instance the machines were actually more of a shield than a Mole. Instead of a cutting-wheel, a back-hoe type excavator was mounted inside the shield. These units were custom designed and built for this one of a kind project, and were quite effective.

Contractor for the tunnel was Guy F. Atkinson of California. The machines were built by the Milwaukie Boiler Co.

Part 4

ESTIMATING

Having spent almost fifteen years (spread over three different periods) in the estimating side of the business, I thought I should recount at least a few of the areas that are perhaps somewhat less common in run-of-the-mill estimating.

TUNNEL ESTIMATING

Preparing a cost-estimate for a tunnel, or other type of underground structure, can be quite different to other types of estimating - above ground structures for example - in both the Building and Civil Engineering disciplines. To elaborate on that, I suggest: The former categories generally entail a considerable amount of "Quantity Take-off". In fact quantities are the basis of any estimate. If the quantity summary is not correct the estimate cannot be correct! Of course, that is a very broad-brush statement.

Personally I have had some experience in two vastly different systems, a) The British system, which is "supported" by a highly regarded professional organization called "Institution of Quantity Surveyors". The "Q.S's" generally prepared all building type estimates - in very minute detail - and applied "Unit Costs" to each item. When you think about it, the smaller the units, the better the chance of having a reasonably accurate estimate for the "whole". However, this only holds true for certain types of construction - and tunneling is not necessarily one of them! I'm just trying to emphasize the point - don't lose track of the main item by spending too much time on details.

Back to the world of tunneling. The difference here is the work is (a) Underground (b) for the most part, a linear project, and (c) We are dealing with a major variable - the Geology throughout the project. There is just no way we can apply "fixed units" as a means of estimating.*

* except in the case of making a "Ball park" - or as it used to be called a "Horseback" - estimate to give management a somewhat close idea of the value of the work.

In underground structures however, such as powerhouses, subway stations etc., each project is so unique that detailed estimating needs to be the norm. No doubt there are a number of computer programs out there these days that will handle a lot of the estimate's features, but I personally would not care to work with them. Having said that I must admit I worked for ten years, in Ohio, for a tunneling company where we used a large computer program. However, there was one huge difference. All of the quantity take-off was done the "old fashioned" way. Crew sizes, equipment costs, materials, etc. were all inserted by the estimator. The difference was, Master Sheets containing Unit Costs for all of the above were made at the outset, by the Chief Estimator. (In the case of Equipment, only "Hourly Operating Costs" are inserted at this point. Ownership Charges and Rentals will be applied separately).

Once the Master Rates are determined, computer codes (mnemonics) are assigned to each individual item of labor, material and equipment. Estimators will then prepare a Work Sheet(S) for each individual bid item and the computer will take care of all the extensions, summaries etc. A real benefit is the fact that the computer can then summarize all types of information such as "Total amount of Class A concrete" in the entire estimate. For a large project, where such an item will almost certainly be adjusted just before bid time, this is a great feature. Man-hrs. for all trades will also be summarized so that if a certain rate was revised since starting the estimate an overall adjustment can be made very quickly.

I again have to admit that I am quoting from many years ago. I'm sure there are some very good estimating programs out there for "Construction and civil Engineering" I still believe however that the only type of program I would recommend for underground work is the type mentioned above, and even that would not be necessary for small projects.

For my own occasional estimating I devised a simple "Excel" sheet which I find works well for me.

However, I caution - do not even use an Excel sheet for estimating unless you have some "fail-safe" cells built-in!

The trick is to ALWAYS have a SECOND "Total" line at the bottom of the sheet, i.e. You may have five vertical columns which you will total down to your bottom line. Your horizontal lines will total across to a sixth column. Insert a cell *below* the total in the sixth vertical column. Formulate this cell to total the *horizontal totals* across. This total of course should match the *vertical total* in the sixth column, or something has been missed out!

I once visited a Consultant's office where the owner showed me the Excel sheet (program) his estimators were using and - believe it or not - on the sheet he handed me I spotted an error in the total - by eye-balling it! They did not have a "check across line". The poor guy was not only astounded but embarrassed! It may have been the only incorrect sheet in hundreds (maybe he gave it to me as a test - I don't think so!)

Drill & Blast: A major factor, in drill and blast tunnel estimating is "Overbreak", especially if the tunnel will have a concrete lining installed. The amount of overbreak will vary considerably, depending on two factors. (1) The type of rock and (2) The blast pattern used, particularly the perimeter drilling and type of powder used. At this point I think I should emphasize that I believe a Tunnel Estimator must have practical experience in *tunnel construction,* ideally as a Supervisor, Surveyor etc. As such, this person has had the opportunity to observe - and hopefully keep notes on day-to-day operations. It is next to impossible for one person to gain experience in ALL of the various types of tunnel construction, but exposure to a few types gives the "feel" of the business of underground work, the people that make it happen, the unexpected things that *always* happen, and most importantly, how to overcome such "happenings".

Of course, making the switch from Field to Office is not always easy - maybe never, but things happen and situations change.

Always make notes on:

1. Crew sizes;
2. Equipment - don't forget the Portal equipment;
3. For drill & Blast: powder used and quantity per "round"
4. General description of ground type;
5. Shift hrs. & number of shifts; and
6. "Bull Gang" size, and operations performed.

There are a myriad of other items that might fall in the category of "time-study", but spot checking is also good for things like rock-bolting, setting steel ribs etc.

Project Overhead

Speaking from my own experience, having worked for a few different companies in "estimating" and having participated in a large number of joint-venture estimates, I believe I can state that most large companies view the category of Project Overhead in very similar ways. The overhead portion of a large estimate will be a quite large percentage of the overall bid. Pre-printed sheets listing all of the usual items for inclusion in site overhead are used. For example all site office-staff, general superintendents, safety supervisor, etc. will be in the "Supervision" section of these sheets. Supervisor's vehicles, all office supplies and all site set-up (office, warehouse, dry-house, garage, etc.) are also included. The Overhead Total is then distributed to the bid items as a percentage factor (O.H./Direct Cost)

Equipment Ownership and Profit Margin is then distributed. At that point we have a "final estimate". However, even this will be adjusted in the closing hours before bid time, with changes made for last-minute material quotations.

Overseas Estimating

Overseas bidding obviously involves a large amount of different factors. However, such factors again, break down into two main phases - Direct, and Indirect Costs.

Direct costs: research on the availability of experienced miners, equipment operators, etc. must be made by the estimator(s). Similarly, availability of the secondary type of equipment to be used must be checked. Primary equipment will most likely be shipped in.

Indirect costs will contain, besides the "norm", a number of extra add-ons items. These may include the following:

- Port charges - including possible intermediate port.
- Port charges may *not* include unloading charges for transfer. Such charges are not necessarily included in the basic Port charge.
- Import Duty. This can vary greatly at some ports, and is not always readily available as a "quotation"
- Personnel: Include moving costs for families of required "Ex.Pats". Accommodations and other expenses - which are usually agreed on before personnel make commitments - are calculated.
- Replacing personnel who decide not to stay overseas *
- Housing; In isolated areas, houses for families and a "camp" with cafeteria etc. may have to be constructed.

Above is a basic listing of the overhead costs which will be distributed to the bid items.

* Personnel turnover costs can be quite high. If the original supervisors etc. have not had experience in the specific country in which the project is based, some changes may become necessary!

JOINT - VENTURES

One of the most interesting experiences in estimating work is taking part in Joint-Venture meetings. When a very large project comes out for bid, especially in a specific field such as tunneling, it is quite usual for two or more companies to form a temporary alliance to bid the job. One of the group will be nominated to be the Sponsor.

The sponsoring company's chief estimator will prepare a listing of all major materials to be used in the work, with a unit price shown for each item. Also a listing of all trades is given, with specific wage rates for all classifications within each trade. This ensures that all of the independent estimates are prepared on an "apples to apples" basis. All of these unit prices are referred to as "plug prices" - to be adjusted to actual quotations, just prior to bid submittal, by the sponsor.

The Joint-Venture Meeting: This meeting usually takes place at the sponsor's office, a few days before the actual bid date. Following general discussions, each partner will read out his estimated costs, first by Bid Total, then by Group Total. (Most large projects are broken down into groups, which have been set-up by the sponsor at an early stage of the estimate). All "site-overhead" costs are considered as a separate group. Eventually it gets down to the major individual *bid items*, these will be examined and discussed in detail.

Eventually a mutually agreed *cost* is hammered-out (although at this point *method* is not necessarily agreed on). The company principles will then have their enclave, and settle on "mark-up" percentage, and the bid amount is determined.

Of course any last minute quotations from subs/suppliers, if lower than prices used to this point, will be taken into account.

EXAMPLE:

OROVILLE DAM. In the early sixties I was working for the McNamara Corp., Toronto. When the Oroville Dam project came out for bid, McNamara joined with a group "Utah Construction", a major Civil Engineering contractor. I had figured the "By-pass Tunnels", which were included in the same contract, so I got to go to the JV meeting at the office of the sponsor - in San Francisco. The meeting was extremely interesting - especially for me.

As dirt-moving was the single major factor of the bid, the method of moving the material from the "borrow" areas to the dam-site was the main subject of discussion at the JV meeting. Some *eighty million cubic yards* of fill material had to be hauled from the borrow pits to the dam location. The maximum haul distance was about twelve miles. I would have to think that the location of such an amount of suitable fill material may have been the deciding factor in the final location of the dam. The JV partners had come up with different methods of moving the dirt. Two of the group figured using 50 C.Y. Scrapers, two selected to go with a twelve mile long, 60" wide conveyor, loaded by front-end loaders at various intervals along the line. However, a ***cost*** was agreed on and the estimate was finalized. Our JV bid of *one hundred and twenty-two million dollars* was submitted. We were second, the low bid was two million dollars less - we should have "rounded-off"!

CONTINGENCIES

A very important and sometimes misunderstood item, in finalizing an estimate, is the "contingency".

Contingencies really apply to *every* project, of no matter what type. It is very important that the estimator(s) and management are on the same page when it comes to discussing contingencies. We all know what the word implies when used in general terms, but in bidding a large project it is essential that some definition, applicable to the project at hand, is set down.

For example: Suppose we are bidding a tunnel project of 10,000 L.F. (3,050M) The usual "Geological Report" will contain bore hole logs and samples of rock taken along the tunnel route at let's say, average of 500 ft. (170M) Centers. Every boring shows excellent quality sedimentary limestone with extremely thin partings. The design calls for minimum supports such as 4 ft. rock bolts in conjunction with flat steel straps, (the TARP project for instance). Now, the first question is "does the contract contain a Changed Conditions clause"? If the answer is "no", then contractor beware! This leads to a moment of decision - do we include a contingency amount in our bid? The estimator should make an educated guess on this and prepare a cost estimate to cover a hypothetical delay. This "side estimate" should be presented to Management who will make a decision as to whether to carry the contingency cost (or add to it) or not.

While using TARP* as an example, I can mention a section where conditions were indicated as just about perfect for machine boring. This WAS the case for quite a few thousand feet. However, within the next thousand feet or so, six clay seams of 6 to 8" thick were encountered. Time lost through that section was about eight days. In a stretch of at least ten thousand feet, that becomes a relatively small delay, but the effect on profit may be a much larger percentage.

* This article does NOT suggest that the TARP projects did or did not contain Changed Condition clauses.

SCHEDULING

Another very important part of estimating is preparing schedules for the project being bid. A simple bar-chart should be part of any estimate no matter how small the project.

However, when it comes to large intricate jobs a Critical Path Schedule (CPM) is a must. So many elements of cost are directly related to project time. All of the "direct costs" are covered in the bid items of course, but Jobsite Overhead and Equipment costs are two of the main factors that are usually left until a schedule is available, as each is directly time-related.

Overhead is based on the *total time* estimated, but equipment requirements usually need to be determined by reference to schedule. Computer programs such as a CPM system may produce total hrs. of specific machines, but I have found that a bar chart may be easier for plotting equipment types, determining overlapping of same, and making decisions as to renting or purchasing. I can remember back in the sixties, flying from Toronto to Vancouver to bid the "Mica Underground Power project" while waiting in the airport lounge I was completing a schedule for just that reason. A consultant had prepared the main part of the estimate for us but had not had time to determine actual numbers of pieces of certain types of equipment (air-tracs & trucks for example). Of course we would have a couple of days to finalize the estimate in Vancouver, and this was pretty much "par for the course".

Just to clarify the matter of equipment in an estimate. The reference above to "numbers of pieces" may be a little confusing to those who are not familiar with preparing large dollar size estimates for projects requiring a year or more to complete. Any large estimate will likely include a need for company owned equipment, plus some short term rentals. As to the former, major equipment, such as a TBM will most likely be charged to the project by a percentage of replacement cost. On really large projects this could actually be 100% of value. Of course it is also likely that such a machine would be custom built for said project, and salvage value would be relatively insignificant. Secondary equipment on the same project, if company owned, would vary considerably in percentage write-off depending on previous usage etc. Such percentages would usually be determined by the company owners. On the other hand, "outside rentals" are simply determined by time requirement (use the schedule) at a quoted rental rate.

Summary: Rental equipment should be carried in "Direct" cost and Major equipment listed with values, presented to management and distributed to the appropriate bid items once management has decided on the "write-off" percentage.

SOME SPECIFIC EXAMPLES

CHICAGO --- TARP. 1977 "ALL-IN" BID

The Chicago Metro Sewer District, in 1976 had decided to proceed with the construction of several miles of large diameter sewer tunnels, and some additional pump stations, to provide a system of flood relief, to be known as the "Tunnel & Reservoir Plan" - TARP. This was a huge undertaking that would involve construction of over a hundred miles of tunnels, ranging in diameter from 18 ft. (5.5m) to 35 ft. (10.7m).

A number of contracts for the larger diameter tunnels were advertised for bidding over the months of May through August, 1976. Three or four of these projects were bid in the ensuing months, but none was actually "awarded". Each of these bids was in the seventy million dollar range. I had been in the U.S. about six months at this point, working as Chief Estimator, for a major tunneling company in Ohio named S & M Constructors. We had been low bidder on one of the above projects (as part of a joint-venture). Following another couple of months, and still no "award" made, it was announced by TARP that a "change in strategy" would be adopted. Instead of proceeding with sections of the overall scheme.

THE TOTAL LENGTH OF THE LARGER DIAMETER TUNNELS WOULD BE ADVERTISED AS TWO ONLY PROJECTS, WITH AN OPTION TO BID BOTH AS A COMBINATION!

This was a staggering move. The combination of the two sections would amount to some twenty miles (32 Km) of tunnel, ranging from 32 to 35 ft. (9.7 to 10.6m) diameter. Also, each contract would include a major amount of "collector sewer" connections, Drop Shafts and

Pump Station enlargements, none of which had been included in the *original* tunnel contracts.

The four company joint-venture group, which S & M had been part of in bidding the now canceled "short" sections, came together and made the major decision to proceed with bidding the new package. Time allowed for preparing and submitting an estimate was twelve weeks. Seems like lots of time, but when the amount of ancillary work is taken into account, the man-hours of quantity calculations and cost estimating that would be required represented a huge cost. The Sponsor for the group would be "Morrison - Knudsen" It was decided that building just one of the two large sections was not very practical, due to limited accesses along the route for work shafts. We would have to bid the "combination" - despite the enormous undertaking.

Note: *The difference between Joint-Venture partners and "Sub-contractors" is that each partner of the J-V prepares a complete estimate, as though he was bidding the Job alone. The sponsor will provide unit prices for all major materials. These prices will be used by all partners so that each estimate is on an "apples to apples" basis. All prices of course will be adjusted by the sponsor, who finalizes the bid, usually in the final twenty-four period before the actual bid time.*

Following a grueling twelve weeks of work, the partners convened at the office of Kenny Construction in Chicago to finalize the estimate. (Normally the bid would be wrapped up at the Sponsor's office, but as the "Kenny" office was on the outskirts of Chicago it just made sense to meet there).

A huge problem surfaced in the final twenty-four hours before bid time. When the (approximate) final dollar amount was suggested to the insurance company - for "Bid bond" purposes - it caused major "scrambling". None of the large companies approached would cover the dollar amount involved! This was a staggering blow. A bid submitted without a Bid Bond would be considered irresponsive.

How much were we talking? I can't disclose the actual number but let me say that a *very minimum* would be $500,000,000. - 1977 dollars!!!

Why were the insurance companies so reticent? After all, we were four reputable companies with substantial resources. For one thing, escalation in that time-period was rampant (particularly in Insurance Costs). Another item, this project was ninety percent underground work - always considered riskier than surface work.

Notwithstanding this set-back, there was nothing to do now but submit our proposal, regardless. A written statement advising that a bid bond could not be obtained at time of submittal, was attached.

As I was not an employee of the sponsoring company I was not part of the actual "close-out" of the estimate. I have no doubt there was a lot heavy sweating in those final hours, mainly, at that point, for the company principals.

Bid opening was set for noon, April 20, 1977. Location, the offices of the M.S.D. Chicago. Our delegation alone was quite a large group. We had arrived just before noon. With a very few minutes left, we were wondering where the other bidders were. The clock struck twelve. The appointed clerk reached under the counter and pulled out a single large envelope. He handed the package to the chief engineer and said "this is the only bid". The Chief said. "One bid only - we cannot open or accept a single bid". The envelope was handed to a J-V representative. Disbelief, sighs all round. That's it, over, kaput!! Three months of hard work, a final week of "pressure cooker" effort, countless thousands of dollars spent. For me, I guess the one consolation was the relief.

For the M.S.D., it was back to "square one". Well, not quite, the total design was in place. That probably would not change. They could go back to the original plan and offer the more reasonably sized projects for bid, (about seventy million dollars each). It turned out that all of the connecting sewer projects were also divided up, accounting for well over a hundred separate contracts. Amazing!

The moral of this story might be - "Don't put all your eggs in one basket". However, it was far too serious a miscalculation, with consequences that set the overall plan back many months, cost for our group alone, a huge amount of money, and tied up a large amount of manpower for a minimum of three months.

Many interesting articles on the TARP projects have been published in the "R.E.T.C. Proceedings" in the 1970's.

The Manapouri Project

In the early 60's I was employed by a large construction company, McNamara Corporation, Toronto, Canada. I was a senior estimator at this point, specializing in the preparation of cost estimates for tunnels and all the associated underground structures - subway stations, pump stations, and hydro-electric stations. There was a large underground Hydro-Electric project getting underway in New Zealand, located near the southern tip of the south island. "McNamara" had contacted a contractor in Wellington, N.Z. and made a tentative arrangement to form a "joint-venture" partnership to bid the Power House portion of the project. To my complete surprise, I was assigned to make the initial trip to N.Z. to meet with the local company and make the mandatory project site tour. A shock to my system - I had been in Canada (From Ireland/England) less than two years, and here I was about to travel across the world, representing a Canadian company !

I remember landing at Wellington airport on a very wet Saturday evening. I made contact early Monday morning, as planned, with our "partner's" representative, an engineer / estimator, named Jim. Flights had been booked for us to Invercargill, a small town located on Lake Manapouri, (the site of the project) with a scheduled stop at Christchurch on the south Island.

As it turned out, our flight ran into some very "bumpy" weather on the approach to Christchurch. We landed without incident (other than a couple of squeamish drops). We proceeded to check on our connection to Invercargill. It seemed that all flights were cancelled indefinitely due to the very unpredictable weather. Jim suggested we could drive the remaining distance and be at Invercargill by late evening.

Well, we rented a car and set out to drive a good portion of the south island. We arrived at the town of Invercargill late evening. We already had a reservation at the local motel.

On schedule the next morning we went to the nearby dock and within minutes a little float plane was landing alongside. My partner and I introduced ourselves to the pilot. We boarded the little plane, and were on our way across the lake. This whole area falls within a national parkland called "Fiordland Park". There are actually two lakes within the park, Manapouri and Lake Te Anua. As we approached the far side of the lake, we could see in the distance, a deep inlet, or fiord, from the Tasman Sea. (This fiord had the ominous name of Doubtful Sound) Our pilot informed us that the discharge of the "tailrace tunnel" would be located at that point. At this time the tailrace tunnel excavation had not yet started, but preparatory work at the outlet end was underway. We landed at a point close to the main field office, where we were met by one of the engineering company's inspectors. Following a briefing by the Senior Inspector, we were shown boxes of "drill cores" - the rock samples taken at various depths in the area of the proposed powerhouse, etc.

Next stop - the "Exploratory" shaft. A small diameter shaft had been excavated to the elevation of the powerhouse, about 200 m. deep. From the bottom of this shaft, two small tunnels about two meters square, were excavated, each reaching about 200 m. out from the shaft, at approximately right angles to each other.

The mine hoist used in the excavation was still in place at the shaft, and Jim and I were lowered down in the "man-basket" to shaft bottom. From there we walked through the excavated length of the two small drifts making independent notes of the rock types, water seepage, etc. Of course we chatted to our guide, who had mentioned that he had worked in the excavation of the drifts. He gave us some idea of the "drillability" of the rock, his estimate as to how often a driller had to change drill bits, how many pounds of powder were used per "round" and so on. This information would be somewhat of a guide, but of course most of the drilling and blasting for the chambers and access tunnels would be with much heavier equipment and larger drills than a "jack-leg" used by a miner. Of course there would also be a lot more exploratory-size drifts required before the bulk work was started.

Back at the main field office, we got more information about the general area, proposed access roads and "camp" location for the powerhouse crews, and contractor's staff quarters. We were given interesting information concerning the facilities in place for the tailrace tunnel. There was in fact no access road at this time. All personnel, supplies etc. etc. were brought in by boat. The "Camp" set-up for ALL personnel was also a boat - well, really quite a large ship!

With some daylight still left, Jim and I were flown back across the lake, this time circling the area where an access road was being cut through the bush to connect the Power House area with the Invercargill side of the lake.

Jim and I got back to Wellington without any problems. I spent a couple of hours at our partner's office the next day, initially comparing notes with Jim, then meeting with his boss and generally discussing the steps to be taken in preparing the estimate. We agreed on the standard steps for closing out the estimate - which of course would take place in Wellington. Although McNamara would be the sponsor, it would be logical for the local company to obtain wage rates for the project and material "plug prices" for use in the preparation of the estimates.

As it turned out, in the ensuing weeks, for various reasons, both parties decided that this project was not one that either wanted to bid! Ah well, a lot of money spent, to "take a look", but that's part of the business.

There are many accounts of the Manapouri Power Project", from inception to final completion, on websites such as "Google", "Yahoo", etc.

In checking out those articles myself, I found some very interesting information concerning the project that I had not been aware of in the ensuing years since visiting the site.

It seems that the expected output of the installed generators was not reached during the initial years of operation. It was resolved that the overall expected dynamics of the system was reduced by the amount of "friction" in the tailrace tunnel! Very simply, the volume of water spinning the turbines was just not getting egress at a fast enough rate, through the "blasted" tunnel (no pun intended!). This in turn would actually reduce the velocity of the turbines by causing a backup. A traffic-jam in a critical area—what a tremendous blow!

Again, in my research on the Web, I learned that a second Tailrace was driven some ten years later, this time using a TBM. This machine would produce a smooth-wall circular tunnel, as opposed to the irregular, rough walls of a drill and blast excavation. It seems that over the first few years of operation a number of serious cave-ins had occurred in the tailrace tunnel, resulting in total plant shut-downs.

The Manapouri Project - (Related Story)

Some ten years after my visit to Manapouri, I was up in Labrador, Canada, managing a tunnel project—still with the McNamara Company. We were using a machine called a Conway Mucker, a rock loader specifically designed for excavation in tunnels. It was not easy to find experienced operators for this equipment, but I was aware of a project that had just been completed out in British Columbia, (several miles of access tunnel, for Granduc Mines). The Conway had been the primary mucking machine used. I asked our home office to do some investigating, and sure enough, they managed to come up with two possibilities. Following some more research I was able to track both of these men down. They were both interested in getting back to tunnel work and making a little more money than their current jobs paid. So, we got two operators!

As I already knew from the phone interview, one of the guys—Stan, was a New Zealander, who had worked - as a Conway operator - on the tailrace tunnel at the Manapouri project. We invited Stan over to the house one evening. While we were chatting about N.Z. I pulled out a photo album and we started looking at the few pictures I had from my trip there. I had snapped a picture of Jim and the pilot standing in front of the little float plane. Stan looked at the picture and said "oh my gosh, I remember him, he was killed when that plane flipped on a trip across the lake". Here we are, the other side of the world, some ten years later - is this an amazing coincidence or what?

Another item that Stan recounted in regard to the tailrace tunnel was the experience of living on the "Camp Ship" in that deep fiord. Precipitation could occur over nine months of the year. (Ironically, the same might be said of Labrador, but instead of rain, it was snow). Not much to do, even if you could go ashore. So, the constant routine was eat, sleep on the ship, get on the train into the tunnel every morning and grind away at the rock for a twelve hour shift. I believe they worked four weeks on, one week off—I'll bet that week was a rush!

Part 5

ENCOUNTERS WITH ADVERSE GROUND CONDITIONS

This section addresses some of the Rock Faults and Soft Ground Adverse Conditions witnessed by the author.

Encounters with Shear Zones in Hard Rock

Pretty much any large underground rock project, tunnel or cavern, will be quite likely to encounter a "fault zone" of one type or another that will most likely result in a time-delay, and of course, a cost-increase. I thought it might be worth mentioning a few that I was witness to, and can therefore recount the methods used to contain the faults and emphasize the "be prepared" motto mentioned in an earlier chapter.

Caverns: Excavation of the two large projects I worked on in Labrador, Churchill Falls and The Iron ore Company Mine, Labrador City, each encountered foliation shear zones at inclined angles of 60 to 70 degrees. (Full geological details of both of these anomalies are recorded in the 1974 issue of "Proceedings" published by the A.S.C.E.)

As to the construction aspect, firstly I will address the Churchill Falls Powerhouse. The fault here was a clay seam dipping at about seventy degrees, a mere six to nine inches thick. There were actually a number of similar seams through the construction area, but most had shown in the exploratory drilling and had been handled with the pattern bolting as excavation progressed. This one however was between the powerhouse and the surge chamber caverns, and was not detected until the Surge Chamber was being excavated, in benches, from the top down.

It's over forty years ago now but I remember the day well. I was underground, as usual, and had a radio call from the project manager, "Shake" Hodgins. "Harry, get up to the office, right now".

I got to my pick-up and drove out the access tunnel. Walking into Shake's office I saw there were two of the consultant's geologists present, and guessed what was at hand. "We need two inch diameter rock bolts, lots of 'em - in a big hurry! Andy will give you details. Have them flown in - charter"

This kind of an order by-passed our warehouse manager, and rock-bolt ordering was one of my little chores anyway. Andy, the geologist gave me quantities, lengths etc. and I was on the phone to our supplier in Toronto.

So, bolts, in lengths of ten, fifteen & twenty feet, were ordered - to be flown in on a C47, making round trips, a.s.a.p.

Installation: Drilling would be done with regular "Air-Trac" drills working off the current bench level and subsequent benches, or temporary rock ramps. Bolts were installed on a five by five pattern. The depth of each hole depended on elevation - the higher up the wall, the longer the bolt.

Bolts: The 2" (5.0 cm) diameter bolts were manufactured by the Williams Bolt Company. Steel was high tensile, knurled outside, similar to reinforcing rod steel. Anchor shells would be provided separately, as the number required would not be the same as the number of bolt sections. Bolts Sections would be coupled in varying lengths to match the depth required. The anchor is expanded by initial turning of the bolt, then the bolt is tensioned - to a torque of sixty kips - by a hydraulic jack which is attached to the threaded projecting end, against a 9" x 9" steel plate. The end nut is then tightened against the steel plate and the jack removed.

Bolts had a hollow center of about a quarter inch diameter. This "hole" was used to pump a very fluid cement grout through. The grout would extrude through the anchor, and under pressure, would fill the small void on the outside of the bolt, from the shell end out to the face of the wall. Apparently the system worked. The plant is still in operation!

The sixty ft. (19m) diam. Shaft, Labrador City.

A very similar situation to that described above. Main differences being the fault seam was a whole lot thicker - up to ten feet in thickness. However, the "dip" angle was steeper, and only some fifty linear feet or so would be exposed in the shaft area, at about ninety feet below surface. The major concern was actually raveling of the soft material in the seam. The simple solution to this was to pour a large block of concrete about 10' x 10' x 60' long at the (vertical) face of the opening. However, excavation of the top ninety feet of the Ore Pass shaft by the "raise" method had to be abandoned as it just was not safe to continue within the limited areas of raises and sub-levels. Working from surface down, we excavated a slot of approx. thirty feet wide and sixty feet long which gave us the ability to contain the wedge by bolting progressively, on a 5' x 5' pattern, and installing the concrete block at the 90 to 100 ft. depth.

Ground Failure: Condition Encountered in Shale

Working as Project Manager for a company named "Schwenger Construction" (Ontario) on a water intake tunnel just east of Toronto, I experienced a somewhat unusual condition while excavating through medium strength shale under Lake Ontario.

We had excavated the work-shaft from the top of the "Scarborough Bluffs" to a depth of 250 ft., (76m) which took us to a point some thirty feet below lake bottom (Lake Ontario) to the top of the tunnel. The support design called for six ft. long rock bolts, four each installed in the arch every four feet of tunnel advance.

NOTE: In this relatively soft rock the "Shell Anchor" bolts are not recommended as the compressive strength of the rock would not support the compressive strength of the anchor. The type of bolt used was a ¾" dia. steel rod, encapsulated in epoxy resin. This system was quite effective in the horizontally layered shale. A flexible tube (the sausage!) of two-part resin is pushed into the hole ahead of the bolt. The bolt is then spun by the drill, mixing the resin in place, and extruding it to encapsulate the full length of the bolt. Hardening is almost immediate and torque is checked before moving ahead.

The tunnel was being driven with a 13 ft. (3.9m) diameter Jarva TBM. At a point about two thousand feet from the shaft (1,500 ft out under the lake) we started experiencing a slight problem in the tunnel arch. As the rock was exposed by the advancing TBM a thin crack started showing at top center, behind the "Tail Can" of the machine!

As bolts were normally inserted within four feet of the shield, it was a simple matter to reduce this to two feet and push ahead for "a short distance"! A minute crack was still evident but no slaking or chipping of the shale occurred. However, we still had to halt mining and have the geologist check out the situation - we *were* under some forty feet of water, with maybe thirty feet of shale rock between the lake bed and the top of the tunnel.

The next operation was to install "Extensometers" at various points through the area of tunnel where the crack existed. After this brief delay we re-commenced mining, with bolts at two foot centers, and the crack became barely visible. The conclusion of the geologists was that the compressive strength of the shale was "very slightly" less than the vertical pressure of rock/water! The difference was more than compensated by the doubling-up of the bolts. Simply put, the tunnel was egg-shaping horizontally by an almost immeasurable amount.

Soft Ground Problem

Now this may be sounding like the proverbial "broken record", but a couple of projects later I found myself in an almost identical situation! (Can it be the luck of the Irish?)

This was a soft ground project - yes there is *very* little rock in the Houston area. However, the "compression" factor was so akin to the Toronto project it was almost eerie.

We had precast, on site, a large number of concrete segments for primary tunnel support. The first batch to be used would be installed at the lower (outlet) end of the tunnel - the point *all* tunnel excavation is started - unless there's some physical factor that precludes same.

This also happened to be the deepest (by a small margin) and the wettest point of the whole 15,000 ft. (4,573m) long project.

A sixty foot deep shaft was excavated, using steel sheet piling. Following extensive dewatering, the tunnel "eye" was cut in the sheeting, and the 13 ft. diameter "Lovat" mining machine was pushed ahead. This machine has a full-circle tail shield which supports the ground while the primary supports (the segments) are installed. I was down in the shaft and saw the first few sets actually come clear of the shield. Things were looking good. These rings were taking weight now and everything looked o.k. "See you later Bob" I called and headed up the ladder.

It was maybe two hours later I went back to the shaft area. The Superintendent, Bob, was at the shaft bottom and called up "take a look at this Harry" I didn't need further prompting, gauging Bob's look of concern. I climbed down the ladders and headed into the tunnel area. No obvious sign of damage. Bob is just ahead of me and is pointing up to the top center. There's a fine crack, right on center! I took a close look at the exposed segments. I turned to Bob, "Get everybody out - NOW". There is only twenty feet exposed between the shield and the shaft sheeting, but that's enough for a disaster. The concrete segments are failing. The circular tunnel section which is generally in compression, is now taking tensional stress in the crown as the top segment is tending to "flatten out".

Everyone is outside the tunnel in very short time. We have an option. "Bob, take some guys up to the yard. I'll send the boom-truck over. Let's get about four sets of those steel ribs loaded and down here pronto, oh, and let's get at least six of those long trench jacks" It would be quicker to wedge the jacks inside the segments before trying to set ribs. Setting two jacks, one each side of the crack, at the mid-point of each of the exposed segment would provide adequate support while the steel ribs were being installed. We had lots of these long jacks on site as we had used them in our "open-cut" section.

It took just about two shifts to gets the ribs secured and to lay some temporary decking. Of course the face doors of the machine had been shut tightly since we started the repair operation. Groundwater pumping had continued to ensure minimum possible loading conditions. We also installed "crack monitors" on each of the damaged segments so that any finite changes could be tracked.

Luckily everything remained stable and we did not even have a sign of water seepage. What we did have was a work stoppage at the mining end - the critical point of the job!

Of course as soon as the crews got started installing the steel sets, I was on the phone to the designer giving him the news!

John was a tunnel design specialist, highly respected in the business and had done consulting work for the Morrison-Knudson Company on other projects. I answered all of John's questions with regard to our manufacture of the segments. "Yes, we were sure that the concrete used in those segments had attained the specified strength. Test cylinders crushed at thirty days after the casting date, were above the 5,000 p.s.i. specified. Everything was carefully monitored and recorded. I.D. numbers were stenciled on each segment as soon as it came from the mold. The reinforcing, as specified, was a sheet of 4" x 4" wire mesh, so there was no chance that steel had been shorted.

John: "let me run some numbers, I'll get back to you in an hour or so".

John calls back as promised. "You will need half inch bars at 9" centers, and we should stick with that probably for the first two to three thousand feet of tunnel".

I'm thinking, not too good - tying steel, time consuming, plus we've got tons of mesh on site already, most of it cut to size.

Hold the phone! I thought of my situation in Toronto.

"Hey John, what about we just double-up, use two layers of mesh, every segment"

"O.K. might work. Let me take a look, will call you in a while"

Now even I can do the simple calculations to figure the comparative cross-sectional area of the two alternatives, but of course John has to check other factors of influence.

Another hour (or two) John calls:
"Looks like the two layers will work. I will fax you confirmation including a sketch showing details".

"Great, we'll start casting tomorrow morning".

We found a small storage yard about a mile away and began moving the "original" segments from our main yard to that location. This ensured that the wrong segments would not be used by mistake when we started back mining. Everyone was confident there would be no problem even with the single mesh after some four hundred feet or so. Not only would we have less cover but the soil would be less sandy and therefore drier. We completed the 15,000 ft. of tunnel without any further problem of segment damage or failure. In fact, I believe we set a world record for length of precast segmental tunnel driven in one shift, 115 L.F.

The previous record was set on the "London Ring Main" water tunnel. However, that may not have been a fair comparison as the Ring Main segments were final liners, whereas ours were primary lining and not as difficult to set. Anyhow, our home office was proud of the accomplishment!

Part 6

PICTURES & SKETCHES

Picture by H.J.Hilton

Churchill Falls: Underground Power House.
Designer: "Acres Canadian & Bechtel Eng."

Picture shows Power House Chamber at an early
stage of setting concrete forms for Generator Bays etc.

Upper level beams and travelling crane are actually permanent equipment - in use by contractor as per specifications. Lower set of rails are contractor's equipment.

Other permanent item: The suspended aluminum ceiling. Again this item was, by specification, available for contractor's use. This was an excellent arrangement as the system provided much better lighting than temporary systems would have provided, both financially and safety-wise.

Original photo by: H.J. Hilton

For the "Westheimer Sewer Project" in Houston, Texas, (Morrison Knudson Const.) we manufactured 15,000 + concrete segments on site. Four of these segments would make a full circle. As the segments were four feet wide, each four pieces covered four L.F. of tunnel. The pieces were 6" thick, reinforced with wire mesh. Although located in a busy urban area, our main work site was some four acres, adequate space for such an operation. The segmental ring was the *primary support* for the soft-ground tunnel. Traditionally Ribs & Lagging would have been the type of support used. However, it was decided we would go with the cast-on-site segments. This turned-out to be a more economical and easier to install system than the traditional steel ribs and lagging might have been, at least in this case.

Ore Pocket at Cormey Mine

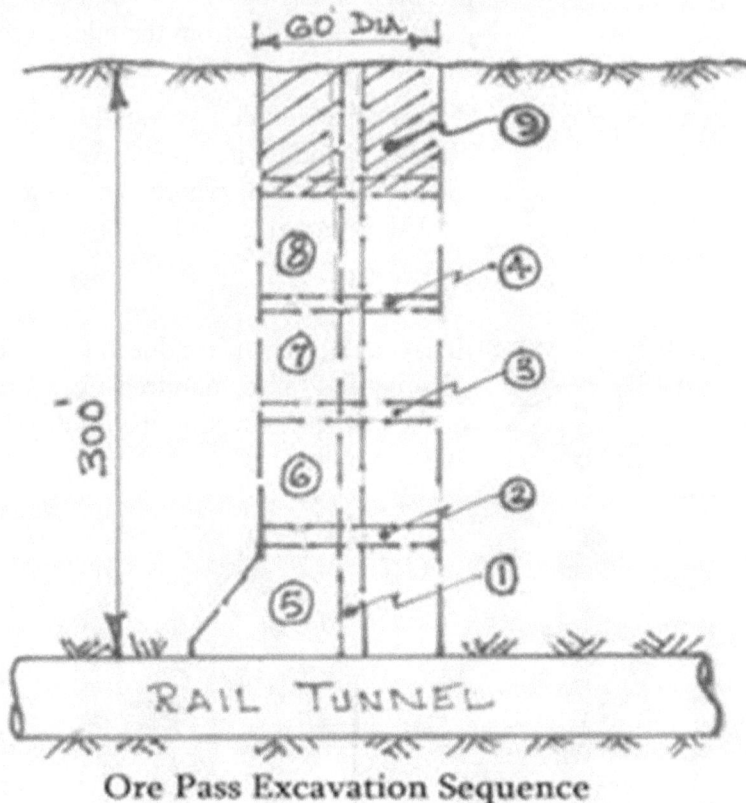

Ore Pass Excavation Sequence

Drawn: H.J.Hilton By. 29. 1969

On the "Iron Ore Company" rail tunnel project in Labrador the lower 200 V.F. of excavation for the Ore Pass shaft was a pure mining-type excavation (as opposed to tunneling). This was sub-contracted to a mining specialist from Montreal, named Masse Gautier. Sections 1 thru' 8 were excavated from the rail tunnel, *upward,* using the "Raise" method.

<u>Section 1</u>	8' x 8' Pilot Raise, approx. 200 ft high
<u>Sections 2, 3 & 4</u>	8ft. high "cross-cuts" - 60 ft. in diameter, excavated laterally from the pilot raise.
<u>Sections 5, 6, 7 & 8</u>	"Sub-levels", drilled vertically from the cross-cuts and blasted one by one to drop to the rail tunnel, where the "muck" is loaded into rail cars, by conveyor, and hauled away.
<u>Section 9</u>	Excavated from surface due to Fault Seam encountered at a hundred feet below surface when excavating the Pilot Raise.

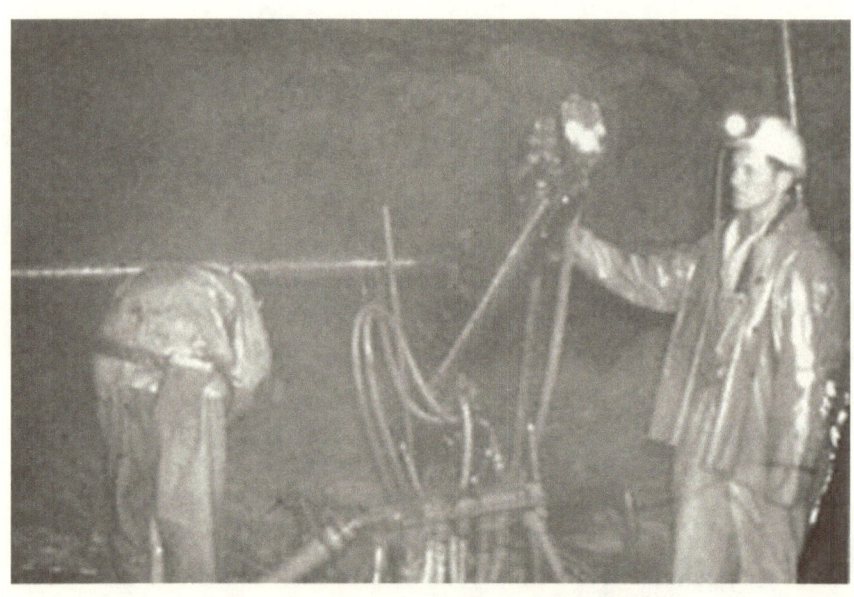

Orig. photo by H.J. Hilton

Miners setting-up for vertical drilling at cross-cut No. 3.

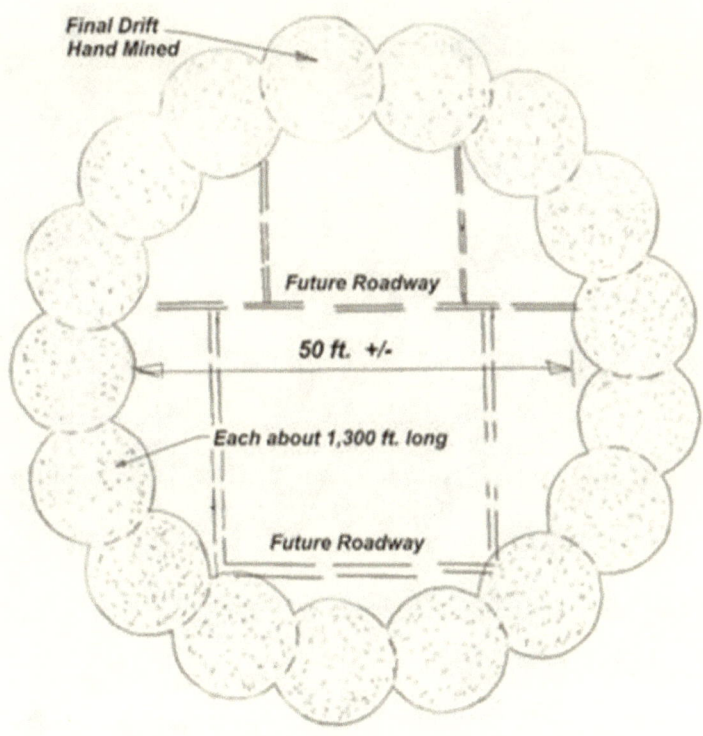

Mount Baker Ridge Highway Tunnel

Sketch, by H.J.H., is purely diagrammatic - prepared for estimating purposes only.

(Detailed information would have to be obtained from the Design Engineer)

A minimum of twenty-four drifts was specified. Each drift had to be filled with concrete before next drift was started.

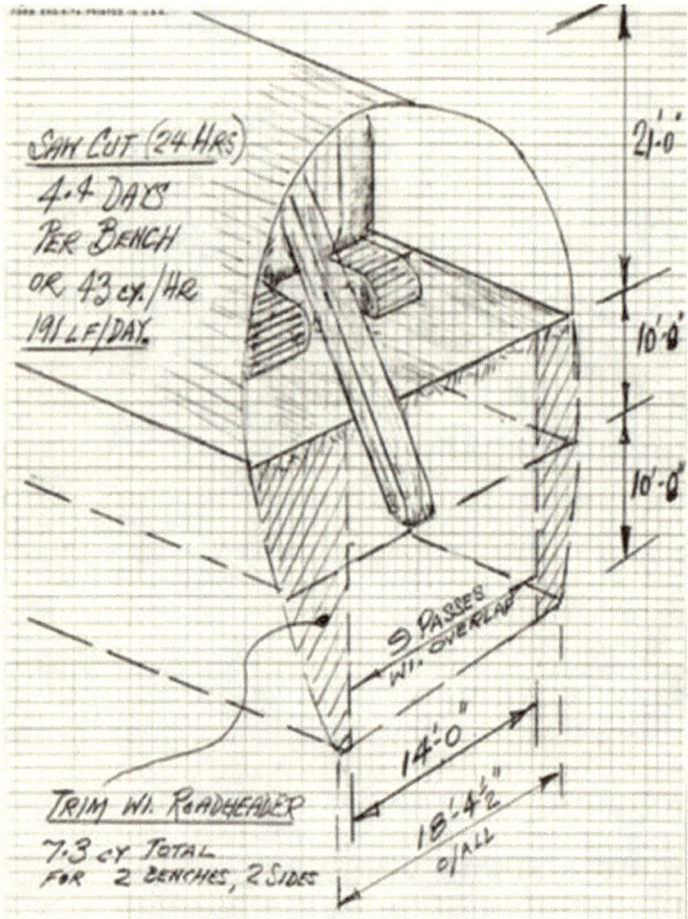

<u>Saw Cutter & Roadheader Combination</u>
<u>(The Roadheader Machine is not shown)</u>

Above is a sketch I made as part of an estimate I worked on for "Morrison Knudson", Boise, Idaho.

These caverns were built as a part of a flood control system, known as "Cole Park" - near Dallas, Texas, approx. value $50,000,000 in 1990. The caverns were quite short, a boring machine was not feasible. We were not the low bidder. I am not aware of the type of equipment actually used by the contractor.

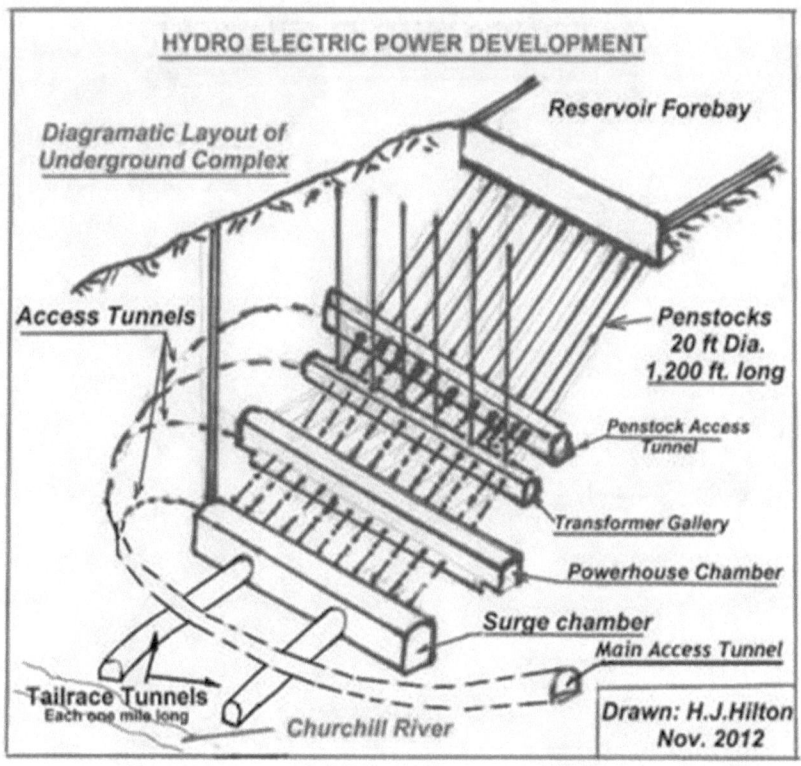

Churchill Falls power Development
Labrador Canada

Diagrammatic layout of the full Underground Complex

Note: This sketch was developed mainly from memory (having made numerous similar sketches while on site) and partly from reference to numerous notes retained from that period.

Picture by H.J.Hilton

Chicago TARP -30 ft. diam. 27,000 l.f. excavated by "Jarva" TBM. A large percentage of this tunnel will not require concrete lining as the rock is very competent. The sections of steel flat-bar in the arch area, pinned by 4 ft. long bolts, are used to prevent "spalling" - caused by initial relaxing of the rock mass. It is also interesting to note that the ventilation line has been removed at this point as there is obviously good through ventilation from nearby shafts.

Part 7

"MURPHY'S LAW" EPISODES

The following incidents occurred on the Labrador City Rail project - Constructed for the Iron Ore Company of Canada.

Contractor: McNamara Corporation of Canada.

"THE SHAKEDOWN"

Labrador City, Canada, 1972

It was a cold evening in September, but then this was Labrador, where it could be snowing in July or August. This vast wasteland of snow, rocks, caribou and moose, stretching from the westerly border of Quebec to the North Atlantic Ocean, is part of the province Newfoundland. Virtually uninhabited except for the coastal town of Goose Bay, which is home for an American Air Force base, a few small mining towns inland, and the Churchill Falls power development

One of those mining towns, Labrador City, happened to be the locale of a tunnel project, and this was the reason this group of tunnel stiffs was there. The Iron Ore Company of Canada had awarded McNamara Construction of Toronto, a ten million dollar contract to excavate a rail tunnel through one of the hilly areas on the mine property. The tunnel would provide relatively cheap rail haulage of the ore from the open pit areas to the smelter plant. The existing rail tunnel could not handle the tonnage being mined now, and truck haulage over the rough hilly terrain, most of the year in sub-zero temperatures, was expensive. The new haulage tunnel would actually dead-end inside the mountain at a point central to the ore body now being mined. Blasted surface rock will be pushed into a three hundred feet deep shaft, which will connect with the tunnel. At that point the rock will be loaded by conveyor into hundred-ton rail cars to start the seven-mile haul to the plant.

Our contract consisted of the excavation, through igneous rock formations, of ten thousand feet of sixteen feet wide by twenty feet high railroad tunnel, about a 1,000 ft. of ninety feet wide "siding", 1,400 ft. of fifteen feet high vehicular access tunnel, and a 300 ft. deep shaft "Ore Pocket", sixty feet in diameter. The mine property was seven miles outside Labrador City, which is located some three hundred air miles North-East of Montreal. Welcome to the cooold!

There are actually two towns, or as they are called, the Twin Cities, Labrador City and Wabash. Both are mining towns. Inhabitants of the Twin Cities are either employees of one of the mining companies or work for the support services necessary to maintain any urban area of some eight thousand people. There was no road into the twin cities. The closest populated area is the city of Sept Isles (Seven Islands), Quebec. Just too far away to make it worthwhile building a road through such territory as lies between. There is a railroad (there has to be some way of getting the ore out) - but passenger service was not available. There is also, an airport at Wabash with reasonably good service to Seven Islands and Montreal.

As Project Manager for McNamara Construction, I had headed up from Toronto to Labrador City with just one other person, Dave Martin. Dave was in fact new to the company. I had hired Dave based on his reputation as a top man with Conway Mucker operations, which was our planned excavation system for the rail tunnel excavation.

Our crews were made up, on close to a fifty-fifty split, of Quebecois and Newfoundlanders, all seasoned miners, mechanics, operators and helpers, most with experience of working on one or more of the large underground power projects built in Quebec in recent years, or at the Churchill Falls underground power station in Labrador. We had built up the crews over the past six weeks or so. We were about ready now to start the access tunnel. The rail tunnel required more extensive set-up, including laying almost a thousand feet of track and switches outside the portal, to the dump area.

The Owner had had the foresight to award a small contract for the portal area of the access tunnel. It included the portal development and excavation of about one hundred feet of starter tunnel. This work was being completed while the main contract was bid and awarded, thereby saving set-up time for the main contractor.

On this particular Friday evening the concentration was on getting the access tunnel under way. It just happened the equipment for this tunnel was more readily available. Removal of the rock here would be by rubber-tired loaders, Scooptrams. These specially designed loaders (also called Load-Haul-Dumps) had eight cubic yard buckets and were a very low profile for operating in mines and tunnels. The drill jumbo was also rubber-tired. We had used an old Euclid rock truck, taken the box off, and built a flat deck to mount the three upper drifters drills. Three more were mounted below the deck. These drills were the same size as would be used at the rail tunnel, but here the feed shells were just six feet long. This was to minimize weight, which would be a problem on the seven percent downgrade of this tunnel. We could still drill 12 ft. deep holes by starting with the six-foot drill steel then switching to twelve footers in the same hole.

It was shake-down time. Every piece of equipment would be test-run. From the pump in the lake, supplying drill water to the drills and everything in between. Also, the two 1,200 c.f.m. compressors, and of course the electrical equipment that powered them, the ventilation fans, with the heaters fired-up. At twenty below zero ambient, you had better be blowing heated air into the tunnel. The airlines and all valves to the drills, and of course, the drills themselves were started up. I watched as Jack Rogers, the Superintendent for this tunnel, had each miner work every piston on the jumbo, lift cylinders, swing cylinders, run each drill back and forth on it's feed shell. Shut off, break down the hoses, drain all water. Fire up the old Euclid engine and move the jumbo into the short starter tunnel. Jack called out "O.K. boys, head for the camp, tomorrow mornin' we start bustin' rock." (he may not have actually said that, but I threw it in for effect).

I did not go to the access tunnel first thing next morning. I went to the rail end instead. I figured let Jack get it started without having me look over his shoulder for the first round. Of course I could not hold out for too long and headed up to the access tunnel after an hour or so. As I pulled up I noticed I was not hearing the expected rattle of half a dozen big hammers. Jack was just walking out of the tunnel with Frank, the master mechanic. He had a chuck-adaptor in his hand. The three of us stepped into the dry-house, no point in freezing outside when there was obviously something to talk about. "What's wrong?" I asked, not wanting to say what I feared. "None of these adaptors work, they don't lock in the chucks". "Impossible" I said, knowing full well that Jack would not be wrong about something like that.

"They came *with* the drills from Jones Co. I had ordered a few dozen for delivery with the rest of the drill equipment to avoid any possible problem—*such as this*. This was not an item prone to breakage, one chuck adaptor could out-last many pieces of drill steel. Also, it was the norm to buy those consumables from the dealer rather the manufacturer, just like automobile parts. "They don't work" Jack said again, with maybe a few adjectives thrown in. "Let's go, bring that chuck son-of-a-gun with you" I called as I headed for my truck.

In the Office trailer, which was located near the Rail tunnel portal, I called the Jones Co. dealer in Seven Islands. He was perplexed at the story but finally decided that we must have been sent chuck adaptors that were now obsolete. He told us a slight change had been made some years ago. He had some of the latest ones in stock. He was most surprised that there were still some of the old ones around. "O.K. I replied, but I can tell you we got them, and we need them changed—in a hurry. I'm sending a man on the afternoon flight". I called to Joan, the secretary, (yes even the office staff worked Saturday on a bush job). "Book Bob on the two o'clock flight to Seven Islands, and on the first flight back to Wabash tomorrow morning" - in case there was more than one flight on a Sunday.

I called Bob our warehouse manager into the office "you're headed for Seven Islands on the two o'clock flight. Go to the Jones shop. We need some new chuck adaptors. Take this one with you. This is the *wrong type*. The one we want has a slightly different shank. Check the ones they have there. Make sure they are *not* the same as this. They will crate up a dozen or so. We are booking you on the Sunday morning flight back to Wabash. Steve will meet you at the airport".

Call from Bob, about five p.m. "I've got the chucks. They are crating a dozen up". "Great, now Bob, you're sure they are the right ones, different lug, you've got the one I gave you, right?" A slight hesitation. "Well actually I don't, but I'm sure these are different". "Bob, put Jack Terry on the phone, and get a steel tape from someone." After some minutes of calling out dimensions, it was clear—these pieces of steel were identical to the ones we had now. Calls to Montreal and Toronto followed. Bottom line, it was finally resolved by the Toronto shop that in fact the "old type" of chuck had been installed in our new drills! Panic stations—some quiet words exchanged. Best solution - get a charter plane on the way, with minimum of six new chucks. "O.K. phone me with his E.T.A. Wabash Airport, and make it soon! This was not only pre cell-phone era, but also even pre-fax days. Telexes were o.k. but sometimes awfully slow. Maybe the only difference is, when things are going wrong, you just don't hear about it as fast.

At least we could move the crew to the rail end to help with that set-up for the day or so. There is nothing worse than absolute downtime when you've got a bunch of miners on hand with nothing to do but stay in camp and play cards, or whatever. With no tunnel driven yet there is no maintenance to fill time.

Finally on Monday afternoon I got the phone call. A chartered Cessna would leave Toronto, stop at Montreal to refuel, and had an E.T.A. at Wabash of 8:00 p.m.

I was at the airport around seven. We had not had a call from Montreal, so I could only assume our man was on schedule. The next hour really dragged, but when it got to be eight fifteen or so, I started getting really anxious. After taking to various people, I finally got the shift-manager to check with the tower. I gave them the call letters for the flight. Reply back from tower, "Yes we confirm that E.T.A. but have no contact with that flight yet." A call to the Jones Co. contact in Toronto failed to shed any light on the delay. He did confirm the flight left Montreal on schedule, but at this point Montreal could not make radio contact either.

At about 10:00 p.m. I decided there was no point in hanging around the airport any longer. I had called home and checked if there was a message there. Negative. Now there was a real concern for the pilot's safety. Early next morning at least that fear was removed. I checked with the airport manager and was told that they had radio contact from the pilot and he was en route from Seven Islands. His E.T.A. was 9:20 a.m. Just after that I also received a call from

Toronto. The pilot had called them early this morning and told them he had to put down at Seven Islands late last night due to foggy conditions. He had gotten somewhat off course and did not have enough fuel to attempt the longer run to Wabash. That unfortunate pilot was having a much more serious problem than we were having. He told me later he was really getting desperate when he finally saw some lights below and figured it had to be Seven Islands. Just about that time he managed to make radio contact with the Seven Islands airport. Of course he stayed overnight in Seven Islands. Poor guy, I guess he needed a change of underwear also.

The next morning we blasted our first tunnel round at the Access Tunnel. New chucks installed, the rest was, as they say, "a breeze." After three days of real aggravation - two actual lost time days—we were finally "bustin' rock. The thing about a tunnel job, or any work that is strictly linear, is there is only one number on a daily report that the home office is interested in, "how many feet did they make yesterday". On a project of several thousand linear feet, no other places to go at least at this time, no incidental operations to send the crews to, when you are down, you are down. But now we are moving, two twelve-hour shifts, six days a week—makes for happy miners. On a "camp" job, the miners expect those kind of hours, otherwise it's not worth the isolation. At the end of the shift, it's a quick change and shower in the dry-house, jump on the bus, head for camp. At camp, run for the mess hall, maybe a beer or two at a buddy's bunkhouse later, then hit the hay and dream of going home in a month or two for some family life and some great Newfoundland fishing!

"THE FAN-LINE STORY"

For the Rail Tunnel we were using a thirty-six inch diameter metal line for ventilation. Typically, for tunnel ventilation, the minimum amount of fresh air required at the tunnel face is calculated in cubic feet per minute. A number of factors are considered, such as number of workers and the cross-sectional area of the tunnel, but the prime factor by far is the total horsepower of the diesel equipment in use in the tunnel at any given time. We usually had three twenty-ton diesel locomotives in use. Their combined horsepower times fifty cubic feet per hp. gives us the starting point for fresh air requirements. Add in the other factors at fixed multipliers and we come up with the total volume of fresh air required at the tunnel face to meet regulations. A practical diameter for the ventilation line (fan-line) is selected, in this case, thirty-six inches. Frictional loss over one thousand feet is calculated and we now have the h.p. required for each fan. Of course when you know the most likely available fan size will be 100 h.p. and the desired line size is thirty-six inches, then it's just a matter of running the calculations and 'backing into" the spacing factor to arrive at the number of fans needed to provide the required volume of air at the *end* of the tunnel. It worked out we could use one thousand feet as the distance apart for the fans, this would give us about fifteen percent above the minimum requirement.

Ten thousand feet of thirty-six inch pipe is just a huge amount of bulk—not very practical to ship by rail to Labrador City. So it was a "no brainer" to decide to manufacture the pipe on site. We had actually done this on a McNamara tunnel project in Colombia, S.A. in the early sixties. In fact I had bid that job and had been involved in most of the equipment purchases, including a portable machine for rolling metal fan line on site, which was quite a new concept at that time. The machine would crimp the edge of the twenty gauge, six inch wide steel. The steel came in rolls of about two hundred pounds. The thin strip of steel was fed into the machine from a roller, and came out the other side as a thirty-six inch diameter "spiral" pipe. The tight quarter-inch wide crimped joining of the continuous metal strip makes an airtight pipe. As each twenty foot long piece slid along the trough carrier the second man of a two-man team would flame cut the end square and the pipe was ready for stockpiling. Hauled into the tunnel on a special fan-line car, the pipe is hoisted in place, butted against the preceding pipe and a foot wide metal band coupling is tightened around the joint.

We had driven about four thousand feet of tunnel at the time of the little episode. The drill jumbo was being pushed into the tunnel from the "lay-by" at the two thousand foot point. The two-deck jumbo, laden with nine big drills and booms was a real rattler, clanking along on the thirty-six inch gage track, pushed by a twenty-ton diesel locomotive. Most of the crew sat on the lower. A couple of the hands rode the top deck as spotters for the locomotive operator who could not see ahead of the jumbo.

While trucking through the long curve, the jumbo leaned slightly to the right. Without warning, one of the booms swung a little further out than normal. The feed shell was cantilevered out. At that point it twisted a little further on the knuckle-joint support and rammed the fan-line dead center. Now the bad luck was, it did not puncture the line, but in a "couldn't do it again if I tried" attitude, it just flattened a short section of the metal pipe. The second piece of bad news was that it hit just a short distance, some twenty feet or so, *in front* of one of the fans. That's when the real show started. The next fan in the line (No. 3) was almost a thousand feet ahead of the flattened spot. It took probably less than a minute for No. 3 fan to pull a vacuum all the way from the dented area at No. 2, to the back of No.3. "You would not believe how fast that line was collapsing- right before our eyes". I think those were the words the Shifter was calling out as I stood looking in disbelief at the long stretch of flat fan-line hanging from the wall hooks. Atmospheric pressure had flattened the twenty gauge metal pipe just as easily as a big man's hand would crush a beer can.

Almost a thousand feet of line had been destroyed. As there was now no positive connection to the outside heater, or to the outside air, operations at the tunnel "face" had to be halted. All hands were manned for removal of the "pancaked" fan-line. We sent in a train of empty muck cars and the crew started hauling the flattened metal with ropes, chains, whatever was available, up onto the cars. There was not enough room to operate even a small loader alongside the cars. It took a full day, two twelve-hour shifts to get all the material out and install the new section of line. Luckily we had enough pipe in our stockpile to fill the long gap.

The next day I called the home office in Toronto and had one of the equipment guys do some research on fanline vacuum protection. He found there was in fact a device available that offered just such protection. A type of trap door that could be installed in the line just behind a fan. The door was balanced, so that if the line further back became blocked, the sudden loss of air in the line would cause the door to drop open allowing an alternate source of air-flow to the fan. This of course would prevent a vacuum occurring in the line. All fans could then be shut down and the restriction removed from the line. Those units cost about $500.00 each. We bought one for each fan. Of course, we never had another incidence of the line being flattened.

"THE CATENARY LINE"

The Iron Ore Company would use electrified locomotives for hauling the ore cars through the tunnel. Power for the locomotives would be through an overhead cable called a "Catenary Line" - similar to the power lines used by trolley cars. In this case the line would be suspended from steel brackets which were bolted to the rock wall of the tunnel at approximately one hundred foot centers. The support arm of each bracket contained a porcelain insulator. These insulators were custom made in Japan for the mining company.

We had completed all ten thousand feet of the rail tunnel, including the one thousand feet of siding, which was ninety feet wide, located near the far end of the tunnel. The Owner had requested that we now install the permanent rail tracks - as an "extra work" item. Of course we agreed. At this point we were down to just one small crew of miners and a clean-up crew dismantling the construction buildings etc. Actually the mine people would install the track, we would take care of the ballasting etc.

With the track in place the Owner would now conduct a final check throughout the tunnel for tights. A "tight" is something the contractor just hates the thought of. It means that some area of the tunnel perimeter is inside the specified minimum.

Inspections had of course been made as excavation progressed and no tights had been found. Now, however the Mine manager decided a run-through would be made with a template, mounted on a locomotive, to the perimeter of the tunnel. After much discussion it was agreed that this item would also be an "extra" as the contractor had cooperated in installing the track. We would be allowed to use Company equipment and this would greatly assist in speeding-up the operation.

A template made of half inch rebar was fabricated by the Company and mounted on one of their locomotives. The outline of the template was nine inches less than required minimum cross-section. Two inspectors and two of our people would ride the locomotive, which would travel at walking pace, and visually check for the nine-inch clearance. Well, it turned out they found ten locations of tight areas. Maximum depth maybe six inches - all located in the shoulder areas - no real surprise. A six inch tight at about fifteen feet above invert, in a drill and shoot tunnel is not easy to spot!

Jack and I talked it over and decided how we would knock these off - without damaging any of the installed service lines - especially the Catenary line. It would be quicker, cheaper and more positive to just take out more rock, at this point, than the minimum.

The company agreed. They gave us the use of a locomotive, an Air-trac drill and a large portable compressor. We made a couple of curtains, thirty feet long, with double layer of heavy wire mesh. These we would suspend from the rock-bolts at about center line of the tunnel. We would use the special half- inch diameter Perimeter Powder only, for each blast. (Perimeter powder is designed to shear *between* holes causing virtually no over-break).

Also, the minimum "slice" would be about three feet high by ten feet long.

Things were going really well, nine down and one to go! The final, and probably the biggest shot was in the siding area. I looked it over with Jack and we both agreed it looked good. The blast holes were drilled, the curtain was secure, tied at the bottom to the rails, and the blast would be almost ninety feet from the Catenary line.

"O.K. buddy, load her up, I'll head out and let the mine guys know the warning horn will go, thirty minutes or so".

I got in my truck and headed back to the office to make some calls. After a few minutes I hear the warning horn and then the blast. I figure the smoke will be clear in about ten, I'll go back in and see how things look. I'm walking back out to my truck when one of our pick-ups comes screeching in. The driver jumps out, a young miner, he looks panicked. "Harry, the ccccatenary line - she's down".

"What do you mean she's down, how much is down"?

"I don't know. We looked around the corner after the blast and

we could see it falling. Jack sent me out to tell you". My worst nightmare - a domino effect - could it possibly be!!! First thing - check at the portal end. If there had been a domino effect it would have run all the way to the portal. My next thought - call Bev and have her start packing!

It's only two minutes from the office to the portal. Relief - I can see the line is still in place at the portal. But how much is down? Just then Jack calls me on his radio (which was silenced of course when a blast is about to be made). "Oh man, we really got lucky. The Sparkie's scaffold was on the track a few hundred feet back. The Catenary cable was snagged on top of it - it stopped right there".

An hour later Jack and I were sitting in the office, blank stares on our faces. "Now tell me, you son of a gun, did you call Joan before you called me, and tell her to pack?"

"No way buddy, I figured I would just head for the airport and call her from Montreal. I guess we'll take that Sparkie and his mate to the Wabush tonight and get them both pie-eyed".

"That's a deal, but first I've got to see Mr. Mine Manager - unless you want to take my place. "Thanks, but I'll just see you in town."

As it turned out, the mine company had a couple of dozen spare insulators on hand. They were already looking into installing intermediary brackets in case they had another "break" incident. There would have been major commotion if all of the line had fallen. But then, the Mine Management had made the decision to install the insulators before the contractor was totally finished work in the tunnel. Plus we had warned them, and they acknowledged there was no such thing as a guarantee when blasting was involved. Oh, what caused the first bracket to break? That was never really established for sure, but somehow a piece of rock must have found it's way through the mesh and made a direct hit!

The Potentials

I've touched on Power Developments, various types of Tunnels, Anecdotes, and the area of Estimating. However, I would like to conclude by returning to my pet interest - the feasibility study of more unique combinations of TBM mined tunnels and caverns for hydropower development.

Innovative design in the field of underground developments must be aggressively pursued. Multiple small diameter tunnels could be driven quite a distance, at reasonable cost, for inlet/outlet tunnels in areas that were perhaps considered uneconomical due to distances from rivers/lakes.

I am not going to attempt to enumerate "end-use" possibilities of cavern/tunnel combinations, but we have only to consider today's trend of fuel storage, coastal and Inland flooding - to name a couple of the major areas of concern - that it becomes obvious that this special area of design warrants much study.

Engineers and Contractors at times do collaborate on design of a specific project. However, there may be a lot to be gained by working together at the *inception* stage of an idea. For example, instead of designing a project and then leaving it to a contractor to provide a large boring machine for the tunnel, why not design a project around an existing machine(s). For example: Let's say a forty foot diameter T.B.M. is being built for a specific project, duration say two years. A totally unrelated Storage Cavern is "on the drawing boards" in another part of the country. By earmarking the large TBM *before* the first project even starts, a huge amount of money can be saved. In fact, a complete "package" would likely be obtained. *The end-use of these projects usually allows for many variations of the geometry, thereby permitting much flexibility for the designer.*

Disclaimer

The author again emphasizes that details relative to design, where contained herein, are not necessarily accurate, but referenced only to present a general overview of certain types of projects etc. for the reader. On the other hand, *construction* details are generally quite accurate.

The names used for individuals throughout the entire text are not the actual names of those persons, (except where referenced in regard to approving the use of pictures etc). The writer has lost contact with most of the people referenced throughout, and did not wish to cause any possible concern for any of those individuals. The same applies to the name Jones used for the drill manufacturer in the "Shake Down" story.

Articles relating to various projects referenced herein, have been published in a number of "Public" textbooks and trade magazines including publications of the A.M.I.E., the A.M.S.E., "World Tunnelling" (London), to name a few. However, as stated earlier, the contents of this book are based strictly on the personal participation of the writer in the *actual construction* and/or the *Cost Estimation* of the referenced projects.

Harry J. Hilton
Author

About the Author

Harry was born and raised in Dublin, Ireland. On finishing high school, he went to England and was hired as a Trainee Engineer by a large Civil Engineering & Building contractor.

Training initially was at the home office and included drafting, quantity "take-off", and surveying basics. Following a year at the home office, he was assigned to a construction project as an Instrument Man, assisting the Project Engineer. In less than two years he was appointed Field Engineer and a year later was moved to rank of Superintendent. His first assignment in that capacity involved supervision of a large sewer installation in a suburb of London. That project included about three thousand feet of small diameter hand-mined tunnel. A succeeding project also included some tunnel work.

In 1960 Harry immigrated to Canada. His first job interview was with a Civil Engineering company. This company, McNamara Corporation, happened to have a tunneling division and from that point on, it became a "life of tunneling" for Harry, with a number of contractors, and then as a consultant.

Harry took a position in the U.S. in 1976 as Chief Estimator, with a tunneling specialty company - S & M Constructors, Solon, Ohio. In 1986 he moved back into field-work and supervised a variety of tunnel projects, both in rock and soft ground over the next ten years.

Now retired, Harry lives with wife Beverley in Kansas. They spend time visiting their five children and eight grandchildren and now, two Great Grandsons ! throughout the U.S. and Canada.

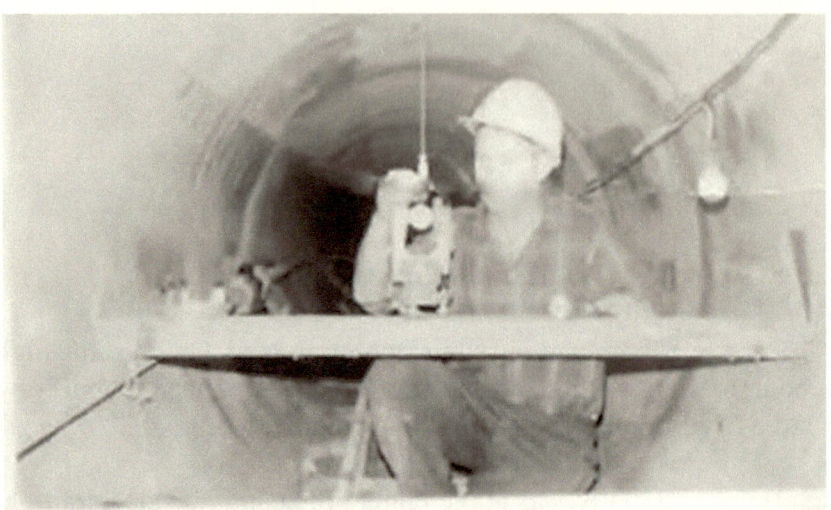

This *very old* picture, circa 1967, shows the author surveying in a 60" I.D. (1.5 m), sewer tunnel, in Toronto, (a little smaller than the tunnel on the front cover!). When there isn't headroom for a tripod, a 2"x8" and a few wedges work just fine. The total tunnel, length 3,000 L.F., (914m) was driven under compressed-air. The formation was a very wet sand, but the application of about five p.s.i. of air pressure dried the sand to the perfect condition for hand mining. The fact that compressed-air was used is also the reason there is no standard ventilation line in the tunnel. Two eight foot long sections were mined and concreted daily (two shifts), by a total *underground* crew of four miners and two "lock-tenders" and two locomotive operators per shift.

*Note: A percentage of net revenues
from this book will be donated
(by the author) to one of the
"Clean Water for Third World Countries"
Active Organizations*

About This Book

Writing this book, for me, has really been a walk down memory lane - the work part of memory lane.

As it is not intended as a technical book per se, I have included a few incidents of on-the-job problems, that I encountered over the years, and how they were handled - figured that's the stuff that doesn't get covered in engineering school!

There is also a section on preparing cost estimates for some major projects, such as the Chicago T.A.R.P. contracts and a few pointers on bidding "overseas" projects. The chapter titled "Murphy's Law" is just that - "It should not have happened" - but it did. A few stories are recounted - just as they happened, about totally unexpected incidents that caused delays, cost money, and caused some minor headaches for yours truly.

Enjoy!